ChemSkill Builder Walk-through

Welcome to Online ChemSkill Builder! This program provides you with the tutorial practice you need to master the concepts covered in your general chemistry course. This incredible online tool contains more than 1500 algorithmically-generated questions, each with tutorial feedback. There is a direct correlation between your time investment in this program and increased problem-solving ability.

Features and benefits of Online ChemSkill Builder include:

- Scores are saved online each time you complete a section.
- Questions are presented in a randomized fashion with a constant mix of variables–no two students receive the same questions. This provides plenty of opportunity for practice.
- Answers are checked immediately, and fully worked out solutions are presented in response to wrong answers. This program teaches you problem-solving skills!
- Twenty-five units include coverage of critical concepts such as stoichiometry, scientific notation, the periodic table, and gases. You will learn problem-solving skills for every topic in the general chemistry course.

ChemSkill Builder

www.chemskillbuilder.com

Student Registration & Walk-through

In order to use Chemskill Builder you must log in. A student login number is included on the inside cover of this walkthrough booklet. Log in as follows.

Registration: Using your browser, go to www.chemskillbuilder.com. Click on "New Student Registration."

Enter your Student Login Number (located on the inside cover) and create your own custom user name. Click on "Submit" when finished.

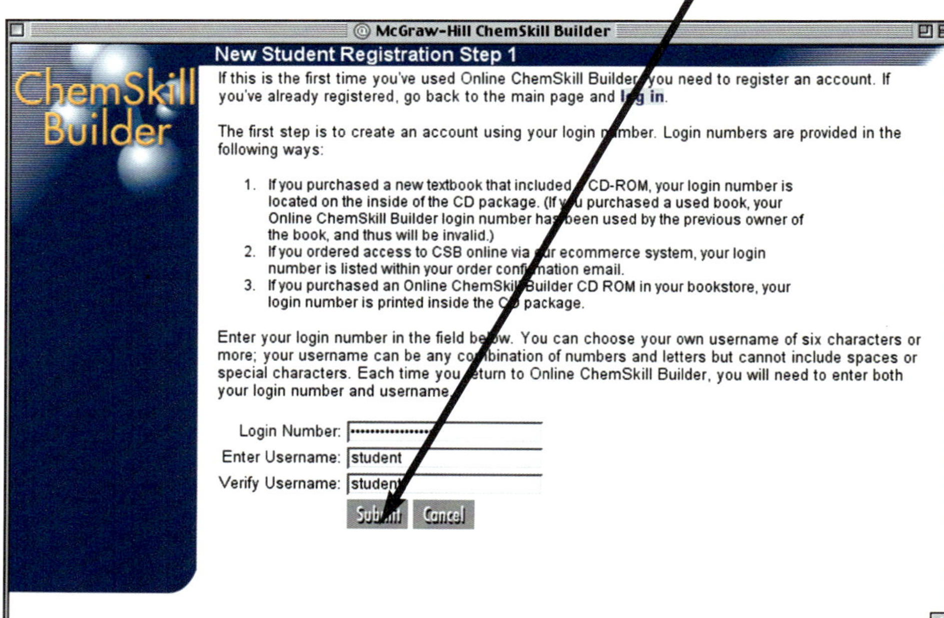

You will need four pieces of information from your instructor to successfully complete the registration process: 1) School; 2) Instructor; 3) Course ID. 4) Section ID. Select the appropriate settings from the various drop down menus and fill in the rest of your personal information. When you are finished, click on "Credit Mode." You will be prompted to verify this information. If the information is correct, click "OK". If not, click "Cancel" and correct it.

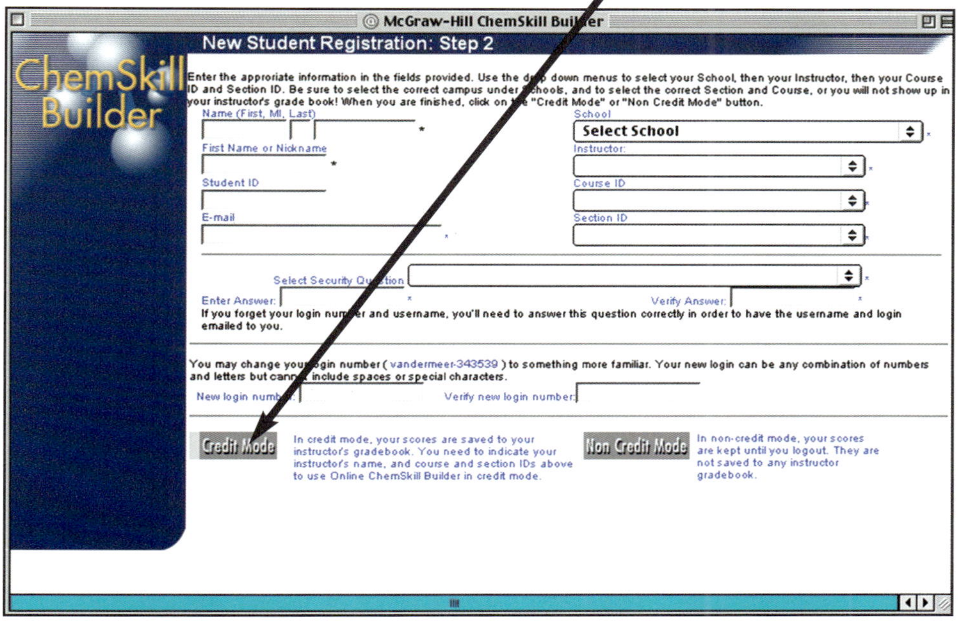

When registration is complete, you will be asked to log back in with your student login number and user name.

You are ready to begin ChemSkill Builder! Click on an assigned unit from the left-hand navigation bar and begin.

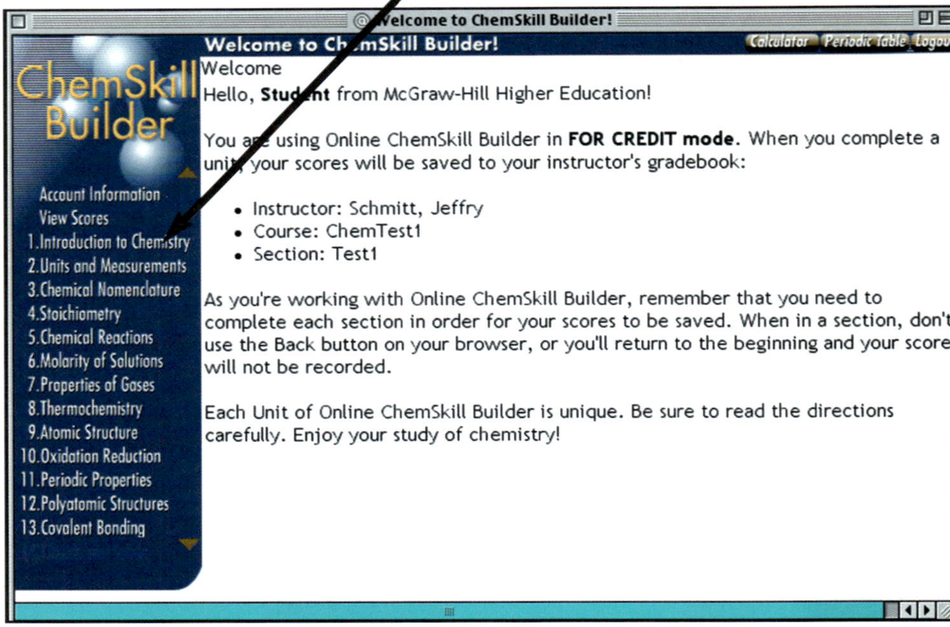

We chose unit 1: Introduction to Chemistry. Each unit contains approximately six sections.

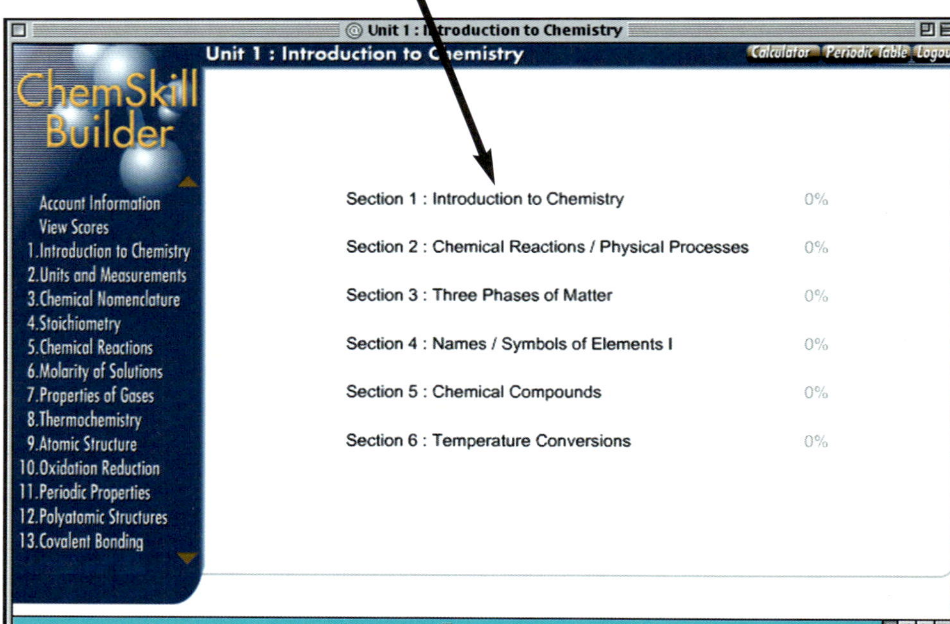

Most sections give a brief introduction on the material to be covered in that particular section.

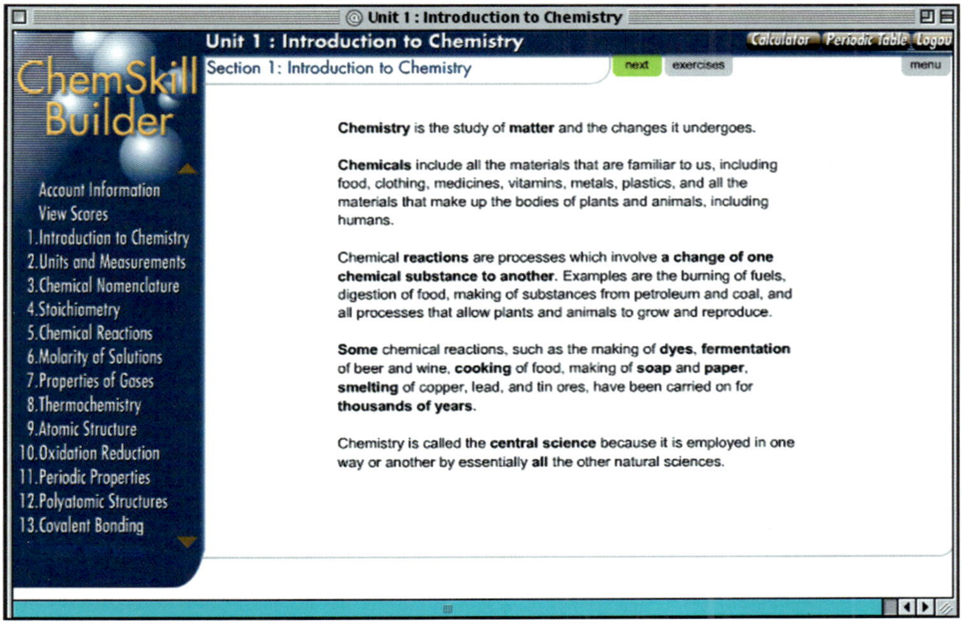

Click on "Next" or "Exercises" to begin your work in a particular section. Before you begin your work, the program will let you know how many exercises are in that particular section. For this "Walk-through" click "Next."

Multiple choice, true/false and fill-in-the-blank questions are offered. Using your mouse, click on an answer and hit the "enter" key on your keyboard.

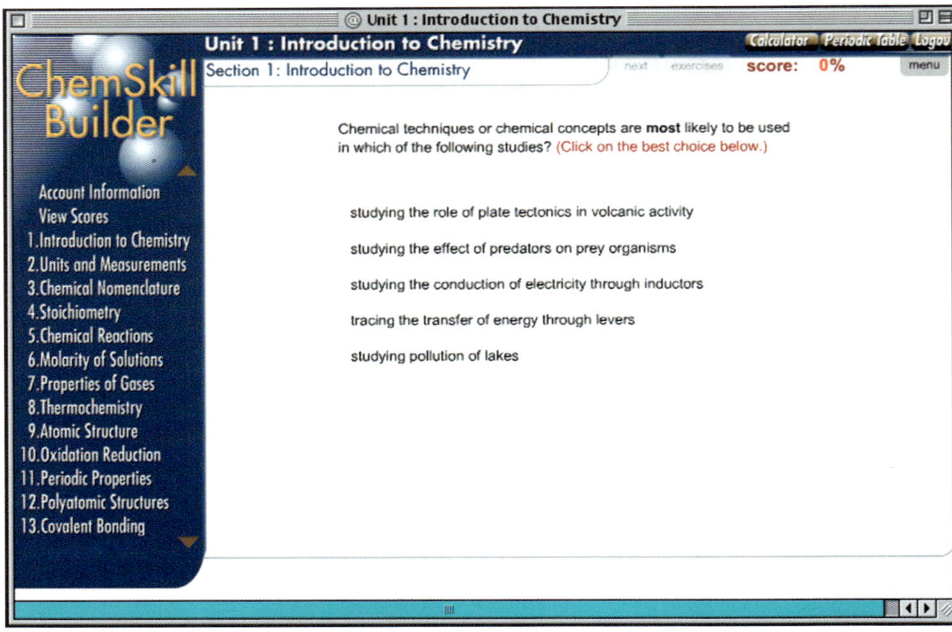

Tutorial feedback is provided when an incorrect answer is given. Scores are tallied at the top of the screen. Partial credit is given if your answer is "close."

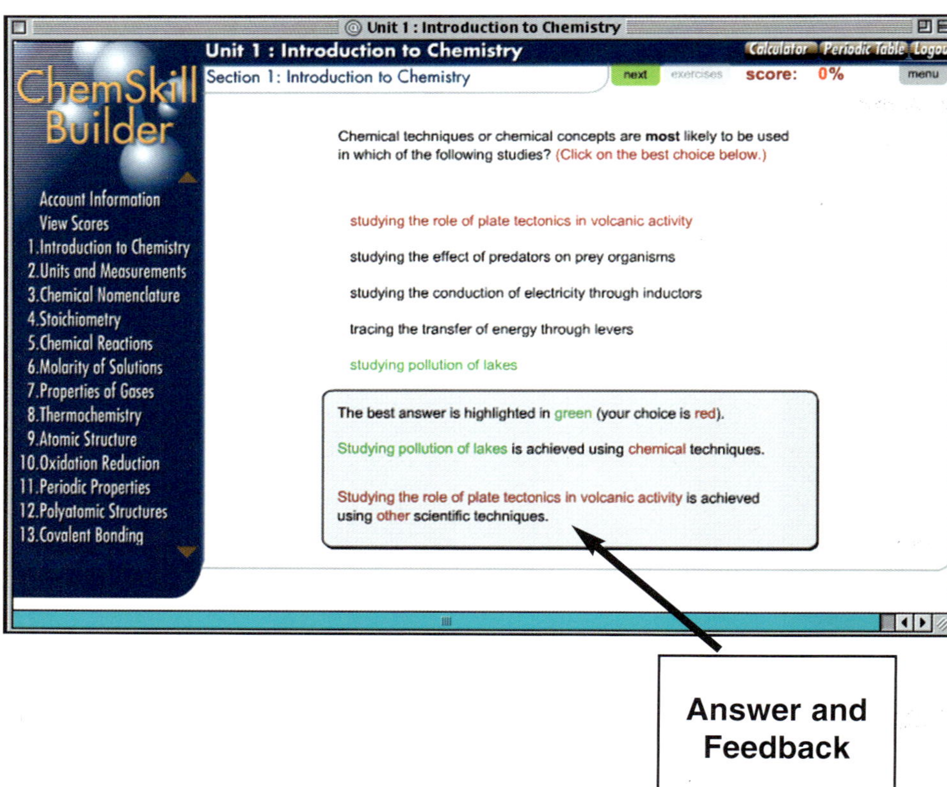

Click "Next" and answer the next question. Continue to answer the rest of the questions in the section.

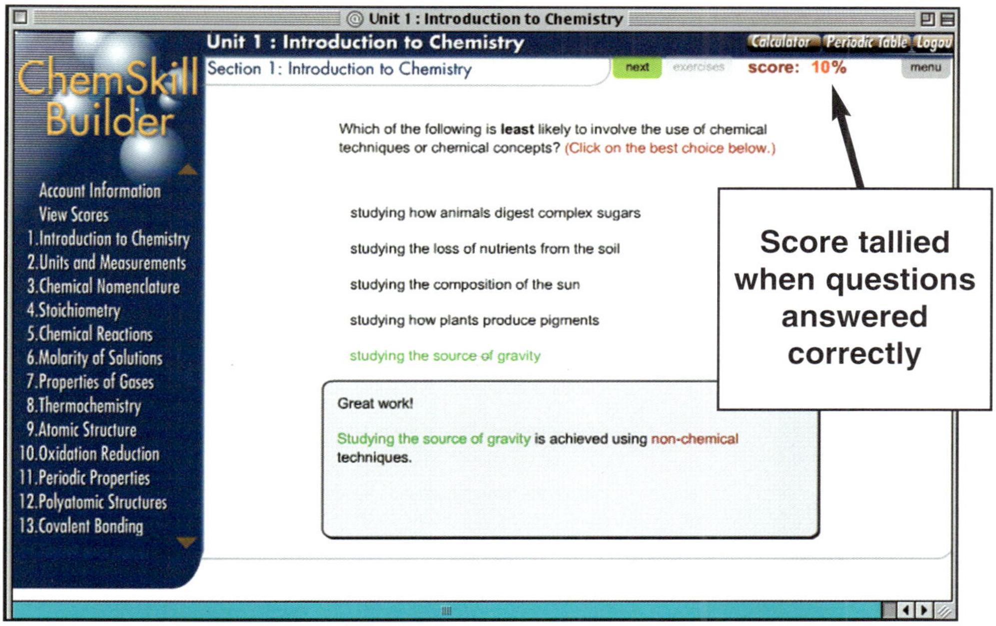

When completed with a section, your scores will save to the instructor's grade book. If you fared poorly on this section, you may go back, do the work and improve your score.

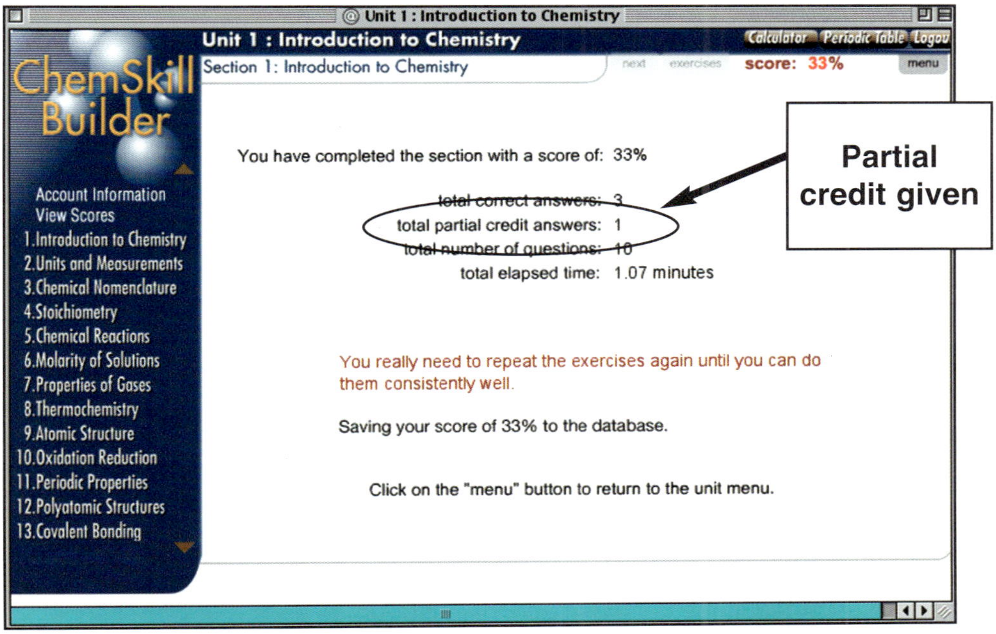

Click on "View Scores" to view scores for all of your completed work in the program.

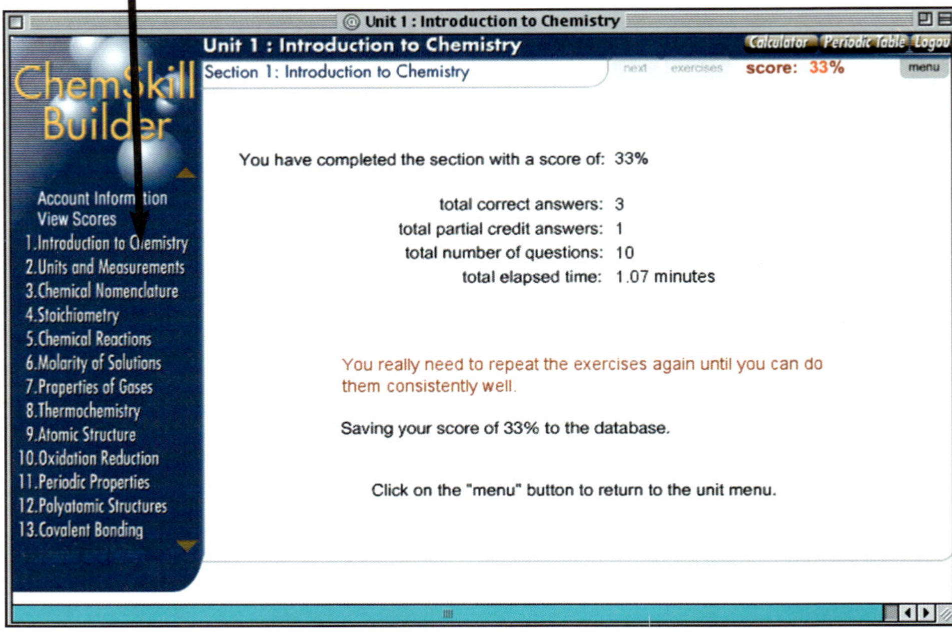

This is how your scores will appear when you view them.

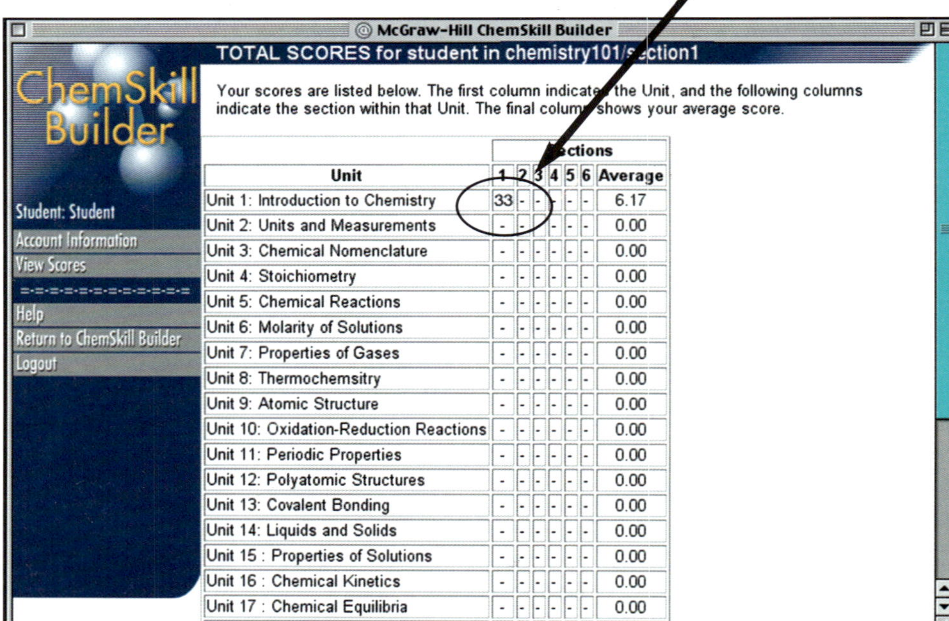

	Sections						
Unit	1	2	3	4	5	6	Average
Unit 1: Introduction to Chemistry	33	-	-	-	-	-	6.17
Unit 2: Units and Measurements	-	-	-	-	-	-	0.00
Unit 3: Chemical Nomenclature	-	-	-	-	-	-	0.00
Unit 4: Stoichiometry	-	-	-	-	-	-	0.00
Unit 5: Chemical Reactions	-	-	-	-	-	-	0.00
Unit 6: Molarity of Solutions	-	-	-	-	-	-	0.00
Unit 7: Properties of Gases	-	-	-	-	-	-	0.00
Unit 8: Thermochemsitry	-	-	-	-	-	-	0.00
Unit 9: Atomic Structure	-	-	-	-	-	-	0.00
Unit 10: Oxidation-Reduction Reactions	-	-	-	-	-	-	0.00
Unit 11: Periodic Properties	-	-	-	-	-	-	0.00
Unit 12: Polyatomic Structures	-	-	-	-	-	-	0.00
Unit 13: Covalent Bonding	-	-	-	-	-	-	0.00
Unit 14: Liquids and Solids	-	-	-	-	-	-	0.00
Unit 15 : Properties of Solutions	-	-	-	-	-	-	0.00
Unit 16 : Chemical Kinetics	-	-	-	-	-	-	0.00
Unit 17 : Chemical Equilibria	-	-	-	-	-	-	0.00

This concludes your walk-through of the ChemSkill Builder Program. Your score has been recorded in the database and instructor grade book. Click on "Logout."

Congratulations, you have now created your student record and have completed a homework assignment.